SCIENTIFIC AMERICAN EDUCATIONAL PUBLISHING

SCIENCE SURPRISES

10 FUN EXPERIMENTS USING CHEMISTRY

BRING SCIENCE HOME

Published in 2024 by The Rosen Publishing Group, Inc.
2544 Clinton Street, Buffalo, NY 14224

Contains material from Scientific American , a division of Springer Nature America, Inc., reprinted by permission as well as original material from The Rosen Publishing Group.

First Edition

Editor: Kristen Rajczak Nelson
Designer: Rachel Rising

Activity on page 5 by Science Buddies (November 16, 2017); page 9 by Science Buddies/Svenja Lohner (November 9, 2017); page 15 by Science Buddies/ Svenja Lohner (August 17, 2017); page 20 by Science Buddies/Svenja Lohner (June 22, 2017); page 25 by Science Buddies/Megan Arnett (May 25, 2017); page 32 by Science Buddies/Megan Arnett (May 4, 2017); page 39 by Science Buddies/Megan Arnett (April 6, 2017); page 44 by Science Buddies/Svenja Lohner (March 2, 2017); page 48 by Science Buddies/Svenja Lohner (December 8, 2016); page 53 by Science Buddies (April 7, 2016).

All illustrations by Continuum Content Solutions

Photo Credits: p. 3, 4, 9, 12, 20, 23, 39, 43, 48, 51, 53, 58 cve iv/Shutterstock.com; p. 4, 5, 9, 15, 20, 25, 32, 39, 44, 48, 53 Anna Frajtova/Shutterstock.com.

Names: Scientific American, Inc.
Title: Science surprises: 10 fun experiments using chemistry / edited by the Scientific American Editors.
Description: New York : Scientific American Educational Publishing, 2024. | Series: Bring science home | Includes glossary and index.
Identifiers: ISBN 9781725349216 (pbk.) | ISBN 9781725349223 (library bound) | ISBN 9781725349230 (ebook)
Subjects: LCSH: Chemistry--Experiments--Juvenile literature. | Science projects--Juvenile literature.
Classification: LCC QD38.S25 2024 | DDC 540.78--dc23

Manufactured in the United States of America

CPSIA Compliance Information: Batch #CSSA24. For further information contact Rosen Publishing, New York, New York at 1-800-237-9932.

CONTENTS

⚛ THESE ACTIVITIES INCLUDE SCIENCE FAIR PROJECT IDEAS.

INTRODUCTION

Chemistry is an important subject to learn in the classroom—and these 10 experiments continue the fun at home! In this book, you will discover cool chemistry processes and activities to further your knowledge of science. Household items, such as Play-Doh and soap, can be used in the activities, as you learn about the way the world around you works!

Projects marked with  include a section called Science Fair Project Ideas. These ideas can help you develop your own original science fair project. Science fair judges tend to reward creative thought and imagination, and it helps if you are really interested in your project. You will also need to follow the scientific method. See page 61 for more information about that.

Make Your Own Magic Dough

SOLID—OR LIQUID? MAKE A MORPHING MATERIAL WITH JUST A COUPLE OF COMMON KITCHEN INGREDIENTS!

Kinetic, or "magic," sand is easy to find in stores, and fun to play with at home! But did you ever wonder what's so magical about magic sand? You might not be surprised to hear the "magic" is actually science! You can create your own version of this fun mixture with just a few kitchen ingredients.

PROJECT TIME

20 to 30 minutes

KEY CONCEPTS

Chemistry
Solids
Liquids
Polymers

BACKGROUND

Kinetic or magic sand is actually just regular sand with one important ingredient added: silicone oil. Silicone is a term used to refer to a group of materials that contain the elements silicon and oxygen. Silicones are polymers, or molecules made up of long chains of repeated units. Silicones are used in everything from hand lotion to car engine oil—they're all around us! The reason we use silicones so prolifically is because they have some very unique properties, including that they can behave as liquids, semisolids, and rubbery solids, depending on the pressure you add. This property is known as viscoelasticity. The degree of viscoelasticity depends on how long the silicone oil's polymer chains are. Longer chains of polymers tend to hold their shapes more effectively than shorter ones do.

In this activity, you'll be exploring the power of polymers by creating and experimenting with your own kinetic dough.

MATERIALS

- 2 cups of flour
- 2 to 3 tablespoons (30 to 44 ml) of cooking oil
- Two bowls
- Fork
- Oil- or gel-based food coloring (optional)

PROCEDURE

- Measure 1 cup of flour into each bowl. Set them side by side so you can make comparisons as you work.
- If using food coloring, add a few drops to your cooking oil and mix.
- Slowly add 1 tablespoon (15 ml) of cooking oil to one bowl of flour. Use your fork to mix the oil into the flour until the oil is no longer visible and there are no big clumps of flour.

- Again, slowly add a second tablespoon of cooking oil to the same bowl of flour. Use your fork to mix the oil into the flour until the oil is no longer visible and there are no big clumps of flour. Use your fingers to blend the flour and oil together. *What do you notice about the texture of the flour? Does it stick together? What if you squeeze the flour in your fist?*

- If the flour does not hold a shape when squeezed, slowly add 1/2 tablespoon (7.4 ml) of oil, mix, and test your flour again. If it still doesn't hold a shape, add the remaining oil. Don't add too much oil! The dough should be dry to the touch. Once the flour is able to hold a shape when squeezed, you have your kinetic dough!

- Put your hands in the oil-free flour. Try to mold a shape in the flour by squeezing it in your fist. *What does the flour feel like? Does it hold a shape when you squeeze it? How quickly does it fall apart? Does the flour stick to your hands and the bowl?*

- Rinse your hands with water and dry them thoroughly. Repeat this step with the kinetic dough. *How does the kinetic dough mixture differ from the flour alone? Can you form shapes with the dough? Do you notice that it sticks more or less to your hands than the flour alone?*

EXTRA

With an adult's help, try molding the kinetic dough into shapes and then slicing it with a knife. Test the dough with clay molds or ice cube trays, and see how well you can get your dough to hold its shape.

OBSERVATIONS AND RESULTS

In this activity, you tested adding oil to flour and observed how this simple ingredient changes the behavior of the flour. In doing so you also created your own kinetic dough. This is similar to commercial products such as kinetic sand or magic sand. What is special about all of these substances is the addition of the oil. The kinetic or magic sand you buy at the store is just regular sand that's been coated with silicone oil. In this activity, the cooking oil you added is a polymer just like silicone oil, and it behaves the same way.

With your kinetic dough, the polymer chains in the oil you added made the flour particles stick together so you can form them into a ball, though it will slowly flatten out over time. In contrast to the flour by itself, you should have found you could form shapes with the kinetic dough and that they would stay together fairly well. You should also have noticed the oil and flour mixture stuck best to itself and not as well to other surfaces. This is because the oil polymers like to grab onto one another more than they like to grab onto things like the countertop. As a result, it should have felt less sticky and easier to clean up.

CLEANUP

Clean off your work surface with a damp towel, and when finished with flour, discard it in trash.

Separate Liquids with Salt!

WANT TO SEPARATE A SOLUTION? JUST ADD SALT—AND SCIENCE!

You probably know some liquids, such as oil and water, do not mix together. If you pour them into the same container, they will form separate liquid layers, one on top of the other. Other liquids, for example rubbing alcohol and water, can be mixed with each other. But did you know that once both of these liquids have mixed you can separate them again into two different layers? How can you do that? The answer might surprise you—with salt! In this activity, you will find out how this works.

PROJECT TIME

20 minutes

KEY CONCEPTS

Chemistry
Solutions
Miscibility
Polarity
Solubility

BACKGROUND

When two liquids can be mixed together, they are "miscible"—they form something called a homogeneous solution, which means that you cannot distinguish the two liquids anymore. In contrast, when they cannot be mixed, they are "immiscible"—they will form two separate layers, called a heterogeneous solution. To be able to mix, the molecules of both liquids have to be able to attract one another. Molecules that are polar (meaning their electric charge is distributed unevenly so they have a more positive side and a more negative side) tend to form hydrogen bonds, whereas nonpolar molecules (which have an equal charge balance) do not tend to form such bonds. Because water molecules are polar, any liquid that does not have polar molecules—such as oil—is usually immiscible with water.

Rubbing alcohol molecules have a polar and nonpolar part, which means they are able to form hydrogen bonds with water and therefore able to mix with it. But how can you break these bonds in order to separate both liquids once they are mixed? You have to add something to the mixture that competes with the alcohol in binding to the water molecules. One substance that can do that is salt. Salt is an ionic compound, meaning it is a substance made up of electrically charged molecules called ions. When ionic compounds dissolve in water, the individual ions separate and get surrounded by water molecules—a process called solvation. Because the salt ions are charged, they dissolve much better in a polar solvent, which is also slightly more charged than a nonpolar solvent. For this reason, salt ions attract the water molecules much more strongly than alcohol molecules do because alcohol is less polar than water. This means that when there is a lot of salt, all the water molecules will bond to the salt ions, leaving none to form hydrogen bonds with the alcohol molecules. As a result, the alcohol becomes immiscible with water and starts to form a separate layer. This process is called "salting out," or "salt-induced phase separation."

Historically, this method has been used in the soap-making process to remove ingredients that should not be in the final soap product. Salting out is also commonly used in biochemistry laboratories to purify proteins, because different protein molecules become immiscible at different concentrations of salt solutions. Chemists use this technique to extract liquids out of a solution, which is what you are going to do in this activity: You will separate a rubbing alcohol and water mixture using just a teaspoon of table salt!

MATERIALS

- Four transparent mini cups (2-ounce size) with lids
- Permanent marker
- Tap water

- Rubbing alcohol (70 percent isopropyl alcohol)
- Table salt
- Set of measuring spoons
- Work area that can tolerate spills
- Ethanol or acetone (can be found in hardware stores) (optional)
- Salt substitute such as potassium chloride or Epsom salt (optional)

PREPARATION

- With the permanent marker label the mini cups 1, 2, 3, and 4.
- Add 1.5 tablespoons (22 ml) of water to cups 1 and 3.
- Add 1.5 tablespoons (22 ml) of rubbing alcohol to cups 2 and 4.

PROCEDURE

- Add 1 teaspoon (5 ml) of salt to the water in cup 1. *What happens to the salt? Does it dissolve in the water?*
- Put on the lid and shake the cup for about 20 to 30 seconds. *What does the mixture look like?*
- Repeat the previous two steps using cup 2 (with rubbing alcohol). *What happens to the salt this time? Does the mixture look different from the water-salt mixture?*
- Take the cap off the permanent marker and swirl its tip in the water in cup 3 for about 10 seconds. Put the lid on the cup and shake it for five seconds. *Does the ink dissolve in the water? What does the solution look like after shaking?*
- Repeat the previous step with cup 4 (rubbing alcohol). *Does the resulting mixture look different? If so, what is different? Can you explain the differences?*

- Next, pour the alcohol from cup 4 into the water in cup 3. Put the lid back on and swirl the mixture for five seconds. *Does the rubbing alcohol mix with the water? What happens to the color of the mixture? Do you see separate layers forming?*

- Now, add 1 teaspoon of salt to the mixture in cup 3. Put the lid on the cup and shake it for 20 to 30 seconds. *What happens when you add the salt to the mixture? Does the mixture look different before and after shaking? If so, how does it look different? Can you explain your results? What color is the mixture?*

EXTRA

Can you separate other liquid mixtures using salt? What about ethanol and water or acetone and water? Try different liquid mixtures to find out!

SCIENCE FAIR IDEA

Are there any other salts—for example potassium chloride, a salt substitute, or Epsom salt—that you could use to separate liquids? Repeat the test, but this time use a different salt than table salt. *Do you still see the same results? If not—how are your results different?*

EXTRA

How much salt do you need to separate the rubbing alcohol and the water? Find out by varying the amounts of salt that you add to the rubbing alcohol and water mixture.

OBSERVATIONS AND RESULTS

You should have seen that the salt easily dissolved in the water in cup 1. (After shaking it the salt seemed to disappear.) Remember that this occurs because the ionic salt molecules easily bond to the polar water molecules. The salt, however, did not dissolve as easily in the rubbing alcohol in cup 2. (Even after shaking it you will still be able to see the salt.) This occurs because the alcohol molecules are less polar than water is, so the salt ions do not bond with them as easily.

With the permanent marker ink you should have observed the exact opposite phenomenon. The ink does not dissolve well in water but it does easily in the alcohol, giving the latter much more color. This is due to the fact rubbing alcohol also has a portion of its molecule that has no charges, and is nonpolar. This portion is more compatible with nonpolar molecules such as the marker ink.

When you mix the rubbing alcohol with water, the alcohol's molecules make hydrogen bonds with the water molecules. The alcohol dissolves in the water to form a homogenous solution, so you cannot distinguish the alcohol and the water anymore. If you add salt to the mixture, however, the salt wants to dissolve in the water and competes with the alcohol for the water molecules. Because there are fewer water molecules available to make hydrogen bonds with the alcohol molecules, the alcohol becomes less soluble in the water-alcohol mixture, eventually forming a separate layer on top of the water. Both layers should have a different color, with the water mostly clear and the alcohol more colored. This occurs because the marker ink is more soluble in the rubbing alcohol.

CLEANUP

Flush all your mixtures down the sink with plenty of cold water. Wash your hands and clean your work area.

Vanishing Baking Soda

CAN A SOLID START TO DISAPPEAR INTO THIN AIR? TRY THIS MYSTERIOUS ACTIVITY—AND SEE WHAT HAPPENS WHEN THINGS HEAT UP!

Baking soda is not only great for cooking, but it is also a useful chemical for science projects. You have probably heard about and maybe even used the baking soda-vinegar reaction to make homemade volcanoes erupt, shoot bottle rockets up in the air, or to detect acids and bases. There are many more chemical reactions, however, that you can explore with the help of baking soda. One of them is called a decomposition reaction, which makes baking soda lose weight! Want to find out how it works?

PROJECT TIME

60 minutes

KEY CONCEPTS

Chemistry
Gases
Chemical reaction
Thermal decomposition

BACKGROUND

Chemicals can undergo many different reactions, such as synthesis reactions that create new substances, or combustion reactions, in which a substance usually reacts with oxygen and generates energy in the form of heat. Another reaction type is the decomposition reaction, which is the opposite of a synthesis reaction. When a chemical substance decomposes, it breaks down to form two or more separate compounds. Most chemical compounds are pretty stable and do not decompose spontaneously. You have to put in a lot of energy to break their chemical bonds. When chemicals are exposed to enough energy, which can be in the form of heat, radiation, electricity or light, they will decompose, however.

Decomposition reactions are the reason why some chemicals or prescription medicines are stored in dark glass bottles. Oftentimes, you do not want a chemical to decompose, because its chemical nature will change. The dark glass reduces the amount of light that reaches the chemical or medicine and therefore prevents light-induced decomposition of the chemical.

Baking soda, or sodium bicarbonate ($NaHCO_3$), is a chemical that can undergo a decomposition reaction when heated. At temperatures above 176°F (80°C), sodium bicarbonate starts to break down into three compounds, forming sodium carbonate (Na_2CO_3), water (H_2O), and carbon dioxide (CO_2). You use this reaction every time you cook and bake. The produced carbon dioxide gas makes baked goods rise! The higher the temperature of the sodium bicarbonate, the faster it will decompose. Only sodium carbonate will remain as a solid product. Both carbon dioxide and water are gaseous at the warm temperature, which means that they will disappear into the surrounding air. You can actually measure this loss—get some baking soda, and find out how in this activity!

MATERIALS

- Two oven-safe plates
- Spoon
- Baking soda
- Digital scale (preferably with 0.1-gram increments)
- Oven
- Oven mitts
- Adult helper
- Heat-safe surface
- Timer
- Paper and pen or pencil

PREPARATION

- Preheat your kitchen oven to 200°F (about 93°C).
- Switch on the scale and zero it.
- Put the first plate onto the scale. Write down its exact mass.
- Then zero the scale again and add 15 grams of baking soda with a spoon.
- Add the mass of the empty plate to the mass of the baking soda to calculate their combined weight. Write down your result on a sheet of paper.
- Repeat these steps with the second plate. Make sure to remember which plate is which. You can label them if you like.

PROCEDURE

- Once the oven reaches its target temperature (200°F, or 93°C), have your adult helper place the first plate with baking soda in the oven. Make a note of which one it was.
- Set your timer to 15 minutes and leave the plate in the oven until the timer goes off. *What do you think will happen to the baking soda in the oven?*
- Have your adult helper carefully take the plate with the baking soda out of the oven. Make sure to wear oven mitts! *How does the baking soda look? Did it change its appearance while being heated?*
- Set the plate onto your work space's heat-safe surface, and let it cool for five to 10 minutes.
- Set your oven to 400°F (204°C).

- Switch your scale on again, zero it, and set the cooled-down plate with the baking soda onto the scale. Write down its exact mass. *How does the mass compare with the mass before you put the plate with baking soda into the oven? Did it get heavier, lighter, or stay the same?*
- When the oven reaches its new target temperature, have your adult helper place the second plate with the baking soda inside.
- Again, set the timer to 15 minutes and leave the baking soda in the oven the whole time. *Do you think you will get a different result at 400°F (204°C)? What will happen to the baking soda this time?*
- After 15 minutes, have your adult helper use oven mitts to carefully take the plate with the baking soda out of the oven. *How does the baking soda look this time? Did it change color or does it still look the same?*
- Set it aside and let the plate cool down for five to 10 minutes.
- Switch on the scale again, zero it, and put the second cooled down plate on the scale. *Write down its exact mass. How does the mass of the baking soda (plus plate) change while being heated at 400°F (204°C) for 15 minutes?*
- Calculate the mass difference of the baking soda before and after the heating at 200°F (93°C) and 400°F (204°C). Subtract the mass of the plate and baking soda after heating from the mass of the plate and baking soda before heating. *How much mass was lost or gained during the heating process? Was there a difference between both temperatures?*
- You can also express the change in baking soda mass by calculating its weight loss as a percentage (the difference of baking soda mass before and after heating, multiplied by 100 and then divided by the initial baking soda mass). *What percentage in mass did the baking soda lose or gain during the heating process at both temperatures? Was it the same? If not, can you explain the difference?*

EXTRA

You can test how the baking soda mass changes at other temperatures. Repeat the same test but set your oven to different temperatures (such as 250°F [121°C], 300°F [149°C], or 350°F [177°C]). *Do you think your results will be different for each temperature or the same? Why?*

EXTRA

Can you find other substances in your kitchen that can decompose? Find out at what temperatures these compounds decompose and, if possible (and harmless), test it yourself!

OBSERVATIONS AND RESULTS

Did you see your baking soda lose weight? You should have—at least for the higher temperature at 400°F (204°C). For the lower temperature, your baking soda mass after heating was probably the same (or almost the same) as before you put it in the oven. Baking soda starts to decompose at temperatures around 176°F (80°C). At these temperatures, however, the decomposition will be relatively slow. Fifteen minutes in the oven at 200°F (93°C) is not enough time to significantly decompose the baking soda.

When you increase the temperature to 400°F (204°C), the decomposition reaction will happen much faster. Fifteen minutes is enough time to decompose the baking soda into sodium carbonate, water, and carbon dioxide. The gaseous products (water and carbon dioxide) will escape into the air, which is why your resulting product should be significantly lighter than what you put into the oven.

The appearance of the baking soda will not change at either temperature. Although you make a new product—sodium carbonate—from baking soda during the decomposition reaction, it will still look the same. Both substances are a white powder.

CLEANUP

Be sure to turn off the oven. Let both plates cool to room temperature. Then, you can dispose of the used baking soda in the trash and wash the plates with soap and water. Wash your hands with warm water and soap too, and clean your work area.

Create Underwater Fireworks with Chemistry

MAKE A COLORFUL DISPLAY UNDERWATER USING THE SIMPLE CHEMISTRY OF OIL AND WATER.

Summertime often brings beautiful fireworks displays. You normally look up into the sky to see fireworks, but in this activity we will take the bursts of color underwater—with chemistry. Although it is not exactly the same as real fireworks, you will be amazed by the color explosions you'll see. Curious about what that looks like? See for yourself in this activity!

PROJECT TIME

20 to 30 minutes

KEY CONCEPTS

Chemistry
Solutions
Miscibility
Density

BACKGROUND

You probably already know oil and water do not mix. But why is that so? Some liquids do mix to become a homogeneous mixture whereas others do not. This depends on their miscibility. Whether a liquid mixes with another depends on their individual molecular structures. Molecules can be classified into polar and nonpolar ones. When atoms come together to form a molecule, they share negatively charged electrons in a chemical bond. Sometimes one atom attracts the electrons more than the other does, which results in a slight separation of the charge into a positive and negative pole within the molecule, also called an electric dipole. When this happens, the molecule is usually polar. Molecules that have an equal charge balance are nonpolar.

A simple rule, "like dissolves like," can tell you whether or not liquids will mix. This means liquids with similar polarity are miscible whereas those with different polarities do not mix. Water is a polar liquid, which means its molecules have electric dipoles. On the other hand, oils are nonpolar, which is why they do not mix well with water. When liquids do not mix, they separate to build two separate layers on top of each other. Which layer is on top depends on the density of each liquid. The density is a measure of mass per unit of volume, which means the heavier liquid will sink to the bottom and the lighter one will float on top. In this activity, you will exploit these properties of liquids to create beautiful color explosions in water that resemble fireworks in the sky!

MATERIALS

- Two transparent glasses (or cups)
- Cooking oil (such as vegetable or canola)
- Water
- Fork
- Spoon
- Food coloring (any color)
- Medicine dropper or pipette
- Two small bowls
- Tablespoon
- A work area that can tolerate food coloring, water, and oil spills

PREPARATION

- Fill one of the cups with room temperature water.
- Fill the other cup with oil.

PROCEDURE

- Into one of the small bowls add 1 tablespoon (15 ml) of oil and 1 tablespoon (15 ml) of food coloring. Mix both liquids with the fork. *How does the liquid look after you mix both solutions? Do both liquids mix?*
- Take the glass of water and place it next to the small bowl. Draw up some of the oil-food-coloring mixture with a medicine dropper or pipette and carefully put one drop into the glass of water. *What happens to the drop when it falls into the water? Does the drop float on top or sink to the bottom of the glass?*
- Add more drops of the mixture to the water until your small bowl is empty. *Does the food coloring change the water's color? How does it look? How long does it take for the water to completely change color?*
- Mix the contents of the water glass with a spoon or fork. *Does the solution look different after mixing?*
- Clean the dropper or pipette.
- Into the second small bowl, add 1 tablespoon (15 ml) of water and 1 tablespoon (15 ml) of food coloring. Again, mix both liquids with a fork. *How does this solution look after mixing? Are both liquids miscible?*
- Take the glass of oil and place it next to the small bowl. Suck up some of the water-food coloring mixture with the cleaned dropper or pipette and carefully put one drop into the oil. *What happens to the drop this time? Does it float or sink?*

- Add more of the water-food coloring liquid to the oil. *Does the oil change color due to the food coloring? How does the mixture look?*
- Finally, mix the contents of the oil glass with a spoon or fork. *How does the solution look different after mixing?*

EXTRA

To create more diverse color explosions, add a range of food coloring with oil in a small bowl. Then mix the food-coloring drops into the oil with a fork. Instead of using a dropper, this time, carefully pour the whole oil-food coloring mixture into the glass of water to create multiple color "explosions" underwater. *What happens with the oil-food coloring mixture in the glass of water?*

SCIENCE FAIR IDEA

Turn this activity into a homemade lava lamp! Use an empty 1- or 2-liter plastic bottle and fill it with about 2 inches (5 cm) of water. Add a couple of drops of food coloring and fill the rest of the bottle with vegetable oil. Then add about one-quarter of an Alka-Seltzer tablet. *What happens to the oil and food coloring when you add the Alka-Seltzer tablet?*

EXTRA

Instead of using water and oil, try to mix food coloring and oil with other liquids. *What do you think happens when you add an oil-food coloring mixture to another liquid such as 50 percent isopropyl alcohol? Does the oil or food coloring mix with the alcohol? Do you see any separation of liquids into layers?*

OBSERVATIONS AND RESULTS

As you probably experienced, oil and food coloring do not mix well. This is because food coloring is a polar liquid, and oil is nonpolar. If you mix the two, you will see lots of little food-coloring drops dispersed in the oil, but the liquids do not mix. When you add this mixture to the water in the glass, again, the oil will not mix with water because this is also a polar substance. The oil will form a separate layer on top of the water—oil is less dense (or lighter) than water. The food coloring, which is also heavier than oil and able to mix with water, sinks to the bottom of the oil drop or layer. As soon as it reaches the oil-water interface, it will start mixing with the water molecules via a process called diffusion.

That means the food-coloring molecules move from a high to a lower concentration of food coloring inside the water. This is why you do not see all of the water changing color immediately, but a slow mixing of both with some parts of the water still clear and others becoming colored. These color bursts and food-coloring trails within the water might have reminded you of fireworks exploding in the sky and slowly falling to the ground. When you mix the solution with a fork or spoon, all of the food-coloring molecules are spread equally within the water and the whole solution becomes colored.

On the other hand, if you mix the food coloring just with the water, both liquids will mix immediately. When you add a drop of this mixture into the glass of oil, you should have noticed the drop sinks all the way to the bottom of the glass, because water is denser than oil. At the same time, the food coloring stays mixed with the water but does not mix with the oil. Even when you mix the solution with a spoon or fork both liquids stay separated. You can use this effect to make your own lava lamp if you did the extra activity!

CLEANUP

You can pour the water and food coloring down the drain. Collect all the used oil in a sealable container (such as a Ziplock bag) and seal it. Then throw the whole container away with your regular garbage.

Shine Bright Like a...Penny!

CLEANING YOUR PENNIES IS A CHEMISTRY ACTIVITY THAT MAKES GOOD CENTS!

PROJECT TIME

60 Minutes

KEY CONCEPTS

Chemistry
Chemical reactions
Acidity
Oxidation

"See a penny, pick it up—all day long you'll have good luck!" Maybe you've heard this phrase before, and maybe you've even stopped to pick up a lucky penny off the sidewalk. But sometimes those pennies you see on the ground look anything but lucky. They appear brown or black, and sometimes they're so dirty looking you can't even tell whether they're pennies! In this activity, we're going to explore why pennies don't stay bright and shiny, and test different methods to bring their shine back!

BACKGROUND

All pennies start out the same color—bright copper. But somewhere along the way, those pennies lose their luster. What happens?

The answer is pretty simple: Although copper is a shiny metal, it is also reactive. The negatively charged oxygen atoms in our air are attracted to the positively charged copper atoms in the penny. When oxygen binds with copper, they form a new molecule known as copper oxide. Copper oxide is brownish or sometimes black in color (depending on other things in the penny's environment). This is why most pennies you see look dirty or tarnished—it's not actually dirt but copper oxide that makes them look so dull.

Luckily, there are a number of ways to clean pennies and make them look bright and shiny again—using chemistry! We're going to test a few different methods for cleaning pennies and determine which gets our pennies the cleanest.

MATERIALS

- Set of measuring spoons
- 10 very tarnished pennies (The more tarnished the better—and the activity works best with pennies that are about equally tarnished.)
- 2 tablespoons of baking soda
- 4 teaspoons of salt
- 2 tablespoons (30 ml) of white vinegar
- 2 tablespoons (30 ml) of freshly squeezed lemon juice
- 2 tablespoons (30 ml) of ketchup
- 2 tablespoons (30 ml) of Tabasco sauce (or other hot sauce)
- 11 plastic cups (Small Tupperware containers or bowls will work as well.)
- Paper
- Water
- Pen or pencil
- Dishwashing gloves
- Permanent marker
- Clear tape
- A timer or clock
- Paper towels
- Access to a sink
- Camera (optional)

PREPARATION

- Make a paste from your baking soda and water: Pour 1/4 cup of baking soda into one of your plastic cups. Slowly add water, 1 tablespoon (15 ml) at time, until you've created a paste.
- Use your permanent marker to label your plastic cups as described below. If using bowls or Tupperware, write your label on clear plastic tape and gently attach it.
- On each of the remaining cups place a label as follows:
    - Ketchup 1
    - Ketchup 5
    - Tabasco 1
    - Tabasco 5
    - Vinegar 1
    - Vinegar 5
    - Lemon 1
    - Lemon 5
    - Control 1
    - Control 5
- Put 1 tablespoon (15 ml) of ketchup into each "Ketchup" cup. Put 1 tablespoon (15 ml) of Tabasco into each "Tabasco" cup. Put 1 tablespoon (15 ml) of vinegar and 1 teaspoon of salt into each "Vinegar" cup. Put 1 tablespoon (15 ml) of lemon juice and 1 teaspoon of salt into each "Lemon" cup. Leave the "Control" cups empty.
- Use your paper and pencil to create a table with three columns and six rows that lists your five test substances (ketchup, Tabasco sauce, vinegar, lemon juice, and control) in a column on the left. Label this column "Cleaning Method." The next column, label "1 Minute." And label the third column "5 Minutes." Make each cell (square) of the table is large enough to place your penny in it after your testing is complete. *Do you have a hypothesis about which substance will best clean a tarnished penny? Why?*

PROCEDURE

- Place a penny on each square of your table, heads side up. If you have a camera, take a picture of the pennies. If you don't have a camera, look closely at the color of each penny. This will allow you to make a better judgment of how well each method cleaned, by looking at the pennies before and after their treatment. *What do you notice about each penny? Can you clearly see the date on the penny or is it too dirty? Can you clearly see Abraham Lincoln's profile?*

- Start with the cups labeled "1." Line these cups up in a row. Take the penny from the "Ketchup, 1 Minute" square of your table, and place it in the "Ketchup 1" cup. Gently swirl the cup so the penny is completely covered with ketchup.

- Start a 1-minute timer.

- Put on your dishwashing gloves.

- When 1 minute has passed, rinse the cup with water until its inside and the penny are no longer covered in ketchup. (Be careful not to lose your penny.) Keep the penny in the cup.

- Repeat this procedure with each of the other pennies in the "1 Minute" column, in turn placing each into their corresponding cup. In each case, make sure the penny is completely covered by the contents of the cup.

- For the "Control, 1 Minute" penny do not add anything to the cup. This is your control. Swirl the penny around the empty cup and let it sit for a minute, just like the other pennies you've tested so far. Rinse it with water.

- Once all of the "1 Minute" pennies have been rinsed, use your fingers to gently scrub the pennies one at a time with your baking soda paste. Keep your gloves on for this part, being careful to keep each penny with its labeled cup. Scrub each penny for 10 seconds, then rinse it again with water and place it back into its cup.

- Set these cups aside for now.
- Repeat these steps with the pennies in your "5 Minutes" column. Start by placing the "Ketchup, 5 Minutes" penny into the cup labeled "Ketchup 5." Again, make sure the penny is completely covered by ketchup.
- Start a timer for 5 minutes.
- Continue to place each penny in the "5 Minutes" column in its corresponding cup.
- Once all the pennies are in their correct cups, use the remaining time to observe the pennies and the contents of the cups. *Do you notice any changes in the appearance of the pennies? What is changing? Do you notice any changes in the contents of the cups? Is anything bubbling or changing color?*
- Once 5 minutes have passed, rinse the cups with water until their insides and the pennies are no longer covered with what was in the cup. (Be careful not to lose your pennies). Keep the pennies in the cups.
- Once all of the pennies have been rinsed, use your fingers to gently scrub the pennies one at a time with your baking soda paste. (Again, keep your gloves on for this part.) Scrub each penny for 10 seconds, then rinse it again with water and place it back into its cup.
- Go back to the table you drew. Use your paper towels to gently dry off each penny then place it in the corresponding square of your table.
- Compare the "1 Minute" and "5 Minutes" pennies in each row. *How does time affect the cleanliness of the pennies? Was 1 or 5 minutes more effective in cleaning the pennies overall? Can you think of why time would be important? Were there any cleaning methods where the time didn't matter, so the "1 Minute" and "5 Minutes" pennies are equally clean? Why do you think this might be?*

- Compare all of the cleaned pennies with the control ones. *Which penny looks the cleanest compared with the control? Which is still the most tarnished?*
- If you took a picture at the beginning, go back and compare each penny's cleanliness before and after being cleaned. *Which penny is the cleanest compared with its before picture? Which is the least clean?*

EXTRA

Test even longer periods of time with each of the cleaning methods. *What is the optimal time to leave the pennies in the cleaning solution?*

EXTRA

Test whether these cleaning solutions work on other types of coins. *What do your results tell you about the special properties of copper pennies?*

EXTRA

Can you think of other safe household acids you could test? What about soda? Orange juice?

OBSERVATIONS AND RESULTS

In this activity, you tested four different cleaning methods to determine which was the most effective at cleaning pennies. You should have observed that each cleaning method made the pennies cleaner but some methods may have been more successful than others. Which method works best is determined by the amount of acid and the presence of salt. Different brands of ketchup and hot sauce will have different amounts of salt and acid in their ingredients, so your results might be different than someone else doing this activity at their house. In each case, however, the cleaning method that should have worked best for you is the one that has the highest concentration of acid.

As they are exposed to the environment, pennies become coated with a layer of copper oxide, making them look tarnished (with a dull, brown, dirty appearance). Copper oxide dissolves in a mixture of weak acid and table salt. Vinegar and lemon juice are both acids. Check the ingredient list for your ketchup and Tabasco sauce—some of the labels might list vinegar or citric acid whereas others might say tomato puree. Tomatoes contain some naturally occurring citric acid, another acid that is generally slightly weaker than vinegar or the amount of citric acid in lemon juice. Therefore, each cleaning method you tested contained acid and salt but the ketchup and Tabasco sauce may have had a slightly weaker acid. In this case, you might have noticed the ketchup and Tabasco pennies were not quite as well cleaned as the pennies that were in the vinegar and lemon juice.

You also might have also noticed the longer the pennies stayed in the cleaning solution, the cleaner-looking they became! The chemical reaction that dissolves the copper oxide on your pennies is an ongoing process. Leaving your pennies in the cleaning solution longer gives the reaction more time to continue, and gives you shinier pennies!

CLEANUP

Discard used supplies, wash and return spoons, and spend your shiny pennies wisely!

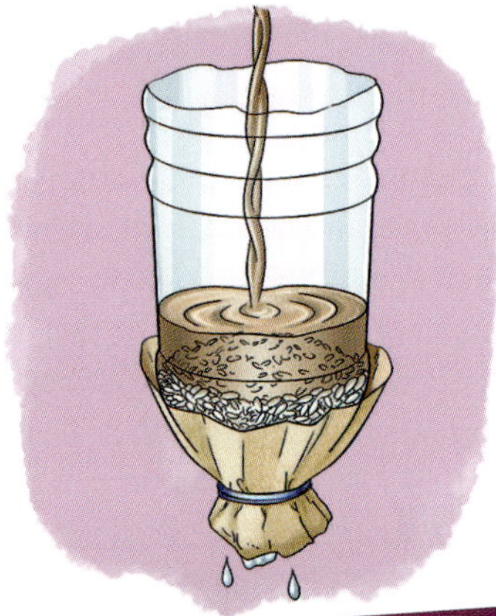

Time to Clean... Your Water!

WHAT GOES ON BEHIND THE TAP? LEARN HOW WATER FILTERS WORK—AND CREATE YOUR OWN!

PROJECT TIME

45 minutes

KEY CONCEPTS

Chemistry
Coagulation
Water filtration

It's easy to take our clean water for granted. We're so used to being able to drink from a fountain or faucet whenever we're thirsty. But did you know more than a billion people in the world don't have access to clean water? Part of the problem is that cleaning (or purifying) water is not an easy task. Not only do you need to remove dirt and other debris, but you also have to get rid of all the invisible bacteria and microorganisms that can make people sick. In this activity, we'll explore a few methods for cleaning water, and learn about the steps involved in making sure you have safe water to drink when you're thirsty.

BACKGROUND

Water purification is a complicated and expensive process. It can be broken down, however, into five basic steps: aeration, coagulation, sedimentation, filtration, and disinfection. The first step, aeration, adds air to the water, simultaneously adding oxygen to the water as it allows other gases to escape. The second step, coagulation, pulls together the dirt and other solid particles in the water. As they're pulled together, the dirt coagulates (forms bigger clumps), which are easier to filter out in the next steps of the process. The third step, sedimentation, allows gravity to work for us. It pulls the heavy clumps of dirt to the bottom of the water, separating it even further from the cleaner water at the surface and getting it ready for the next step—filtration. Filtration involves separating the dirt and other solid particles from the water. Finally (and perhaps most importantly), the fifth step is disinfection. We disinfect water by treating it with chemicals and/or heat to remove bacteria and other unsafe microorganisms that can make us sick.

You will be exploring several of these steps by testing a few water filtration systems of your own. And next time you drink water from a tap, remember how much work went into making sure that water is clean and safe for you to drink!

MATERIALS

- Two paper coffee filters
- Three clear plastic water bottles
- 1 cup uncooked rice or beans
- Five clear plastic or glass cups
- Two sheets of white paper
- A pen or pencil
- 4 cups (950 ml) of water
- 1 cup of loose dirt
- One large plastic container with a lid (for mixing water and dirt)
- Ruler
- Scissors or an X-ACTO knife
- An adult helper
- Rubber band
- Timer or clock
- Measuring cup (at least a 1/2 cup)
- 1/2 teaspoon of alum
- A spoon

PREPARATION

- Combine your water and dirt in the large plastic container. Shake the container to ensure the dirt and water are well mixed.
- Use your ruler and pen to evenly divide one sheet of paper into eight sections, four rows down and two columns across. In column 1, label each row as follows: row 1—"Control"; row 2—"Coffee Filter"; row 3—"Rice and Coffee Filter"; row 4—"Rice, Coffee Filter, Alum" (Note: If you use beans, write "Beans" instead of "Rice" for those labels.)
- Have your adult helper carefully cut the bottoms off your water bottles, leaving the bottles and bottle mouths intact.

PROCEDURE

- Measure 1/2 cup (125 ml) of the dirty water and pour it into one of your plastic cups. Add 1/2 teaspoon of alum to the water and stir gently for three to four minutes. *What do you notice about the water as you stir?*
- Set the cup with alum and dirty water aside; let it sit undisturbed for at least 10 minutes.
- Starting with bottle 1, cover the mouth of the bottle with a coffee filter. Use a rubber band to secure it in place.
- Turn the bottle upside down (so the bottle mouth faces downward). Hold the bottle over one of your clear cups.
- While you hold the bottle, ask your adult helper to carefully pour 1/2 cup (125 ml) of the dirty water into your bottle. Start your timer.

- Hold the bottle over the cup until all of the water is in the cup. In column 2 on your paper, next to the square that says "Coffee Filter" write how long it took for all of the water to get into the cup. *What do you notice about the water in the cup? Does it look different than it did when you poured it in the bottle? What is left in the bottle?*

- Put the cup with water from bottle 1 on your paper, in the square where you wrote the time. Use a measuring cup to measure how much water is in the cup. Write this amount next to where you wrote the time. Pour the water back into the cup and leave it on the paper. *Did all of the water that you poured into the bottle make it into the cup? If not, where do you think it went?*

- Put bottle 1 aside.

- Prepare bottle 2. Cover the mouth of the bottle with a coffee filter. Use a rubber band to secure it in place.

- Turn the bottle upside down (bottle mouth downward). Hold the bottle over one of your clear cups.

- Have your adult helper carefully measure and pour 1/2 cup of rice into bottle 2. Tap the sides of the bottle so that the rice settles at the bottle mouth.

- While you hold the bottle, ask your adult helper to carefully pour 1/2 cup (125 ml) of the dirty water into your bottle. Start your timer.

- Hold the bottle over the cup until the bottle has no water left and all of the water is in the cup. In column 2 on your paper, next to the square that says "Rice and Coffee Filter," write how long it took for all of the water to get into the cup. *What do you notice about the water in the cup? Does it look different than it did when you poured it in the bottle? What is left in the bottle?*

- Put the cup with water from bottle 2 on your paper, in the square where you wrote the time. Use a measuring cup to measure how much water is in the cup. Write this amount next to where you wrote the time. Pour the water back into the cup and leave it on the paper. *Did all of the water that you poured into the bottle make it into the cup? If not, where do you think it went?*

- Put bottle 2 aside.

- Prepare bottle 3. Cover the bottle mouth with a coffee filter. Use a rubber band to secure it in place.

- Turn the bottle upside down (bottle mouth facing downward). Hold the bottle over one of your clear cups.

- Have your adult helper carefully measure and pour 1/2 cup of rice into bottle 3. Tap the sides of the bottle so that the rice settles at the bottle mouth.

- This time, have your adult helper carefully pour the water from the first steps (with the alum) into bottle 3. Try to avoid disturbing any sediment that collected in the cup. Start your timer.

- Hold the bottle over the cup until all of the water is in the cup. In column 2 on your paper, next to the square that says "Rice, Coffee Filter, Alum," write how long it took for all of the water to get into the cup. *What do you notice about the water in the cup? Does it look different than it did when you poured it in the bottle? What is left in the bottle?*

- Put the cup with water from bottle 3 on your paper, in the square where you wrote the time. Use a measuring cup to measure how much water is in the cup. Write this amount next to where you wrote the time. Pour the water back into the cup and leave it on the paper. *Did all of the water that you poured into the bottle make it into the cup? If not, where do you think it went?*

- Measure and pour 1/2 cup (125 ml) of dirty water into your last clear cup. Place this cup on your paper, in the square next to the one that says "Control."

- You should now have four cups lined up on your paper. Look at the water in each cup as well as the time and measurements you wrote on your paper. Compare the results of each of your three tests, with the Control cup. *Which water is the dirtiest? Which is the cleanest? Which filter system took the most time to filter the water? Which took the least? Which filter system absorbed the most water? Which allowed the most water to pass through it?* (**Note:** Although your water may appear clean—it is not, and you should not drink it. Use it to water your yard or plants instead!)

EXTRA

Try this activity with other household materials such as cotton balls, paper towels, sponges, or anything else you think might be a good filter. *Did you find one that worked well? Why do you think it worked well—or not?*

OBSERVATIONS AND RESULTS

In this activity, you tested several different methods for filtering dirty water. Your results may have varied slightly depending on the type of dirt you used in your dirty water. You should have found, however, the water was cleanest when you used the coffee filter, rice, and alum all together. This filter did the best job of cleaning the water because it incorporated several different water purification steps. First, you added alum to cause coagulation. You should have observed the water coagulating as you stirred the mixture during the first step. Next, you allowed the water to sit undisturbed for at least 10 minutes. This resulted in sedimentation, where gravity pulled the coagulated substances to the bottom of the cup, leaving cleaner water at the surface. Finally you filtered the cleaner water through rice and a coffee filter, separating those coagulated chunks from the water.

You should have found the coffee filter alone did a fairly good job of filtering the water, however the extra steps you took with the alum yield better results. In addition, you probably observed that you had the most water come through the coffee filter alone. Both the rice and alum absorb some of the water, so less makes it into your cup from the bottle. The coffee filter alone should have been the fastest filtering method of the three you tested, and have taken the least amount of time to empty the bottle.

CLEANUP

Discard coffee filters and recycle water bottles, and return supplies to their proper locations!

Build the Best Big-Bubble Wand

BIGGER AND BIGGER... HOW BIG CAN YOU MAKE A GIANT BUBBLE BEFORE IT POPS? TRY THIS SIMPLE SCIENCE ACTIVITY AND FIND OUT!

Have you ever tried to blow a bubble, and no matter how hard you try—that bubble just won't form? Why does this happen? Why do bubbles form at all? And why is it harder to blow bubbles sometimes? In this activity, we're going to explore how bubbles form, and test the limits of how big our bubbles can get!

PROJECT TIME

20 to 30 minutes

KEY CONCEPTS

Chemistry
Surface tension
Water molecules

BACKGROUND

As you may know, bubbles rely on surface tension to hold together. Water molecules like to cling to one another and, just like magnets line up with and attach to other magnets they're attracted to, water molecules line up with one another and form hydrogen bonds, creating surface tension. This surface tension allows the surface of water to act like an elastic membrane, stretching and holding a shape. If you drop a small amount of water on a piece of wax paper, you can see a great example of surface tension in action. Instead of splashing or flattening, the water will form small, spherical droplets on the paper. These water droplets can hold their shape because water molecules are more attracted to one another than they are to the wax paper. The strength of that attraction helps hold the water droplet together.

The surface tension of water is so strong in fact, that it prevents us from blowing bubbles from just water. Adding soap or detergent lowers the water's surface tension, allowing bubbles to form. It does this by creating space between the water molecules, decreasing the strength of their attraction to one another. If you think again about magnets, when two magnets are very close to each other, the pull between them is much stronger than when you slowly separate them. Decreasing the strength of the attraction between water molecules lowers the surface tension of the bubble solution, allowing bubbles to form!

MATERIALS

- Two dowel rods (At least 1/2 inch [1.27 cm] diameter works best.)
- Two screw eyes (a closed loop with a threaded base)
- Yarn or baker's twine (at least 18 feet [5.5 m])
- Two washers
- Bubble solution (Optional make-your-own bubble solution ingredients are listed on page 41.)
- A large bucket to hold bubble solution
- Outdoor area with space to run!
- An adult helper
- A measuring tape
- Scissors

- **Bubble Solution** (Makes approximately 1/2 gallon [1.8 l].)
- 8 cups (1.8 l) of water
- 1/2 cup (125 ml) liquid dish soap (Blue Dawn works well.)
- 1 tablespoon (15 ml) glycerin
- 1 tablespoon baking powder

PREPARATION

- This activity could get a little messy. From this point forward everything should be done outside!
- Pour the bubble solution into your bucket. If you're making your own bubble solution, combine all ingredients in your bucket and gently stir to combine. (Try to avoid creating a sudsy foam on the surface of the solution.)
- Ask your adult helper to help you attach a screw eye to one end of each rod.

PROCEDURE

- Cut a 3-foot (0.9 m) length of twine and thread it through one washer and each screw eye, making a loop. Tie the end of the twine together (the knot can be anywhere). This is your bubble wand!
- Set your bubble solution bucket in an open area. Place your measuring tape on the ground, starting at your bucket and extending at least 10 feet (3 m) in the direction the wind is blowing.
- Hold your rods on the ends without the screw eyes. Gently touch the two screw eyes together, and completely submerge the yarn in the bubble solution.
- Gently lift your wand from the bubble solution. Separate the ends of the rods to open the yarn loop.

- Hold the wand away from your body and slowly walk along the measuring tape, in the direction of the wind. *How big of a bubble do you think you will be able to create?*
- Have your adult helper measure the length of the bubble you create! Try it at least five times, and remember the longest.
- Cut the twine and remove it from the wand. Save the washer!
- Cut a 6-foot (1.8 m) length of twine, and thread it through one washer and each screw eye, making a loop. Tie the end of the twine together (the knot can be anywhere).
- Repeat the above bubble-making steps, using the bigger wand. Measure the longest bubble you can create with the 6-foot (1.8 m) twine. *Do you think these bubbles will be easier or more difficult to form?*
- Cut the twine and remove it from the wand. Save the washer!
- Cut a 9-foot (2.7 m) length of twine and thread it through two washers and each screw eye, making a loop. Tie the end of the twine together. (The knot can be anywhere.)
- Have your adult helper assist you while you repeat the bubble-making steps, with both of you holding the wand. Measure the longest bubble you can create with the 9-foot (2.7 m) twine. *Do you think you will be able to create bubbles with the 9-foot (2.7 m) wand? Do you think they could be much larger?*

EXTRA

Test the bubble solution. Try removing the glycerin or the baking powder. *How do these ingredients affect the bubbles you create?*

SCIENCE FAIR IDEA

Let the bubble solution sit overnight, then retry this activity. Test to see if you get bigger bubbles!

OBSERVATIONS AND RESULTS

In this activity, you created some of the biggest bubbles possible, with just a little string and some patience! You may have found that it was easier to create the long bubbles with the shorter twine. As the twine got longer (especially at 9 feet [2.7 m]), it may have been more difficult for the bubbles to form.

We know bubbles can hold their shape as a result of surface tension. In fact, the bubble solution mixture that makes up a bubble's surface is composed of three very thin layers: soap, water, and another layer of soap. This is called a soap film. The surface tension of the water layer is what holds the bubble together. Therefore, when the water between the soap layers evaporates, the bubble pops. To make bigger and stronger bubbles, we need to thicken the walls of soap, to prevent the water from evaporating. To do this, we added glycerin, to help thicken those soap walls and hold the bubble together! This makes for stronger, bigger, and longer-lasting bubbles.

At a certain point, however, bubbles get too big to hold together. Larger bubbles have larger surface area, which means more places for external forces, such as air pressure and gravity, to have an effect. In addition, the larger bubbles have their own weight to support. Imagine building a tunnel with LEGOs. You can make a very long tunnel with a smaller diameter. If you try to make a tunnel that an adult could walk through, however, it's going to have more trouble supporting its own weight. This is similar to what you experienced with the 9-foot (2.7 m) twine wand. The body of the bubble was much larger, and therefore much more fragile.

CLEANUP

Pour the soap mixture down the drain and put away your other materials.

Fire-Fighting Foam

NOW YOU CAN LEARN TO EXTINGUISH FIRE WITH A CHEMICAL REACTION!

Do you have a fire extinguisher at home? Hopefully, yes! It can save your life in case there is a fire. But have you ever wondered how these extinguishers work? Some of them don't even contain water, which is what you probably think of first when it comes to stopping a fire from spreading. What else other than water can extinguish a flame? Do this activity to find out—and put out a candle as if by magic!

PROJECT TIME

20 Minutes

KEY CONCEPTS

Chemistry
Gases
Acids
Reactions
Combustion

BACKGROUND

What actually happens when something is burning? A fire, also called combustion, is the result of a chemical reaction in which oxygen gas reacts with the substance that is burnt. Luckily, not everything catches fire right away when exposed to air or oxygen. For a fire to start, certain requirements have to be met: First, you need a combustible material, or a fuel, that can burn. Second, this material needs to be hot enough so that it can catch fire. The combustion reaction then has to create enough heat to sustain the fire. And lastly, a fire needs oxygen to burn. If either heat, fuel, or oxygen are absent, there will be no fire.

Fire extinguishers are designed to remove one of the necessary "ingredients" for a fire to burn. There are several ways this can be done. For example, water can be used to remove heat from a fire. Adding water often cools down the fuel enough so that it stops burning. Some types of fire extinguishers use dry chemicals to interrupt the chemical combustion reaction. Yet another option is to cover the fire with a layer of foam, carbon dioxide gas, or dry chemicals that stop air from reaching the flames. Each of the aforementioned methods results in the same outcome: The fire is extinguished. You can see for yourself in this activity, and you don't even have to use a real fire extinguisher—just some baking soda and vinegar. Curious about how these two ingredients can put out a flame? Then go ahead and try it out!

MATERIALS

- Small candle
- Matches
- Adult helper
- Vinegar
- Baking soda
- Tablespoon
- Play-Doh or modeling clay
- Glass dish in which to place your candle (The candle should not be too tall; it must be lower than the top of the dish.)
- Water
- A safe area in which to perform this activity

PREPARATION

- Use the Play-Doh or modeling clay to stick the candle into the center of your glass dish.
- Sprinkle a couple of tablespoons of baking soda around the candle. The bottom of the dish should be fully covered with about a quarter inch of baking soda.

PROCEDURE

- With the help of an adult, light the candle with a match. Observe the flame for a couple of seconds. *What does the flame look like? Is it big or small?*
- Carefully pour some vinegar onto the baking soda in the glass dish. It should be enough to make all the baking soda dissolve. Make sure, however, that no liquid or foam reaches the flame. *What happens when you add the vinegar to the baking soda? Can you explain your observations?*
- Once you have added the vinegar, watch the candle carefully. *What happens to the flame? Does the candle keep burning?* **Note:** If you do not see anything happening, try to increase the amount of baking soda or vinegar that you add to your glass dish.
- Try to light the candle with a match again if the flame went out. *Is it easy or hard to do that? What happens to the match if you get close to the candle?*
- Clean out your glass dish and repeat the experiment again, but this time instead of vinegar, pour water on top of the baking soda. *What happens to the flame this time? Do you observe the same reaction? If not, can you explain the difference?*

EXTRA

Can you find other liquids in your kitchen beside vinegar that make carbon dioxide when mixed with baking powder? Remember, these liquids have to be acidic. Observe your burning candle when you add these solutions to the baking soda. *How do they react? Does your flame keep burning, or is it extinguished?*

EXTRA

Together with an adult, locate a fire extinguisher in your house (if you have any) and read its ingredients list. *Can you find out how it works to extinguish a fire?* Ask an adult about how to use this extinguisher in case of an emergency. But be careful to not accidentally use it for real!

OBSERVATIONS AND RESULTS

Did you successfully extinguish your flame? You don't even see what makes the candle go out, but it does—almost like magic! Initially, your candle should have burned bright and steady. After all, it had enough fuel from the candle, heat from the match, and oxygen from the air to sustain a fire. You should have observed, however, that after you added the vinegar to the baking soda in the glass dish, a short time later the flame suddenly went out. What happened? The answer has to do with the chemical reaction that occurs when you mix vinegar with baking soda.

Baking soda is another name for the chemical compound sodium bicarbonate. It reacts with any type of acid such as vinegar to form gaseous carbon dioxide (CO_2). You can actually see it being produced—it makes the bubbles and foam that you observe when you add the vinegar to the glass dish. The same carbon dioxide reaction makes a baking soda volcano erupt or inflates a balloon when you mix an acid with sodium bicarbonate. In this experiment, the CO_2 extinguishes the candle flame. Although you can't see it, the CO_2 produced by the baking soda-vinegar reaction starts filling the glass dish from the bottom up. Eventually, once all of the air in the glass dish is replaced by carbon dioxide, the flame has no more oxygen left to burn and it goes out. If you try to light the candle again, it won't ignite because the match also goes out once it enters the carbon dioxide layer in the glass dish.

When you add just water to the baking powder in the glass dish, however, no carbon dioxide is produced because water is not an acid. Therefore, the candle should have kept burning as long as enough oxygen and fuel was available.

CLEANUP

Make sure you put out the candle. Rinse out the glass dish with water and soap, and discard the Play-Doh or modeling clay.

Battling Gas Bubbles

CAN YOU STOP GAS BUBBLES FROM FORMING—WITH CHEMISTRY?

Has your stomach ever ached so bad you felt as if it was blown up like a balloon—maybe after the big Thanksgiving meal? You may have had a large amount of gas trapped in your stomach and intestines. But where does the gas come from that makes you feel so uncomfortable? And how does it disappear again? In this activity you will blow some bubbles to investigate how a specific compound can help battle those pockets of air.

PROJECT TIME

30 minutes

KEY CONCEPTS

Chemistry
Gas
Surface tension
Human body

BACKGROUND

Everything you eat ends up in your gastrointestinal, or digestive, tract. This includes your stomach, intestines, and other organs that break down the food you eat so that it is useful to your body. Some bubbles of gas (which come from many helpful microorganisms that live inside your gastrointestinal tract to help your body digest food) occur naturally in everyone's body. As gas bubbles form, they can get trapped within the food being digested.

Although a little trapped gas in the gastrointestinal tract is normal, stress or foods with lots of starch can result in more gas production—and large amounts of trapped gas bubbles can cause you to notice it. And the gas needs to escape your body somehow. Some compounds are made to help people expel large amounts of trapped gas. One drug that does this is called simethicone, which is an antifoaming agent. Simethicone does not prevent gas from being made. Instead, it helps the body get rid of the gas at a faster rate than normal.

How does simethicone do this? It has to do with a phenomenon called surface tension, which is created because water molecules cling to one another and pull inward at the water's surface. This is why a water droplet has a round shape. Simethicone reduces the surface tension of the formed gas bubbles, which prevents bubble formation. It also helps to combine lots of small bubbles into larger bubbles, which can be expelled more easily. Want to see for yourself? Don't worry, for this activity you'll just test this on bubbles!

MATERIALS

- Two anti-gas chewable tablets containing simethicone
- Metal spoon
- Sheet of paper
- Two plastic cups or glasses
- Marker
- Water
- Liquid dishwashing detergent
- Two straws

PREPARATION

- Find a work area that can tolerate liquid spills.
- Label the two cups with your marker. On one cup write "WITH SIMETHICONE"; on the other one write "WITHOUT SIMETHICONE."

- Remove two chewable simethicone tablets from their packaging and place them in between a folded sheet of paper. Fold over the edges of the paper on the three open sides.
- Use the back of the spoon to crush the tablets into a powder.
- Carefully put all the powder into the cup that you labeled WITH SIMETHICONE.
- Fill both cups with water until they are about half full.
- Mix the water solution with simethicone using the metal spoon.

PROCEDURE

- Put a clean straw into each cup.
- Slowly blow through the straw that you put in the cup without simethicone for about 10 seconds. *Do you see bubbles form inside of the water? Did bubbles collect on the surface of the water?*
- Next, slowly blow through the straw in the cup with simethicone, again for 10 seconds. *What happens in this cup? Do you see bubbles this time?*
- Now, add about five drops of dishwashing detergent into each of both cups. Stir the solutions with the straws. *How does the liquid in each cup look now? Do both cups look the same or is there a difference? What happens to the liquid in each cup when stirred?*
- Once you have mixed in the liquid detergent, again slowly blow through the straw into the cup without simethicone for about 10 seconds. *Do you see more or fewer bubbles compared with just water without detergent? Are bubbles accumulating on the surface of the water? How many?*
- Repeat the above step but with the cup that contains the simethicone and detergent. *What happens this time? Do any bubbles form on the liquid's surface? Did you expect these results? Can you explain them?*

EXTRA

Try out how much simethicone is needed to be effective. Repeat your tests, but instead of two simethicone tablets, only add one-quarter, one-half, or one crushed tablet to the water. *Do you see the same effects as with two simethicone tablets? Is there a minimum amount of simethicone you need to see less bubble formation?*

EXTRA

In your tests, you used water as a liquid. In your stomach, however, you have lots of acid, called gastric acid. *Does the simethicone also work in an acidic environment?* Repeat the tests, but this time, use an acid such as distilled vinegar (acetic acid) instead of water. *How well does simethicone work in this liquid?*

SCIENCE FAIR IDEA

Can you find other substances besides simethicone that can reduce the surface tension of your bubbles in water with detergent? Do some research on gas relief and surface tension and try different liquids or compounds that you find in your kitchen. *What other substances help reduce bubble formation? Are they more or less effective than simethicone?*

OBSERVATIONS AND RESULTS

You may have noticed that it is pretty hard to make bubbles using only water. In the cup with just water and simethicone, you probably also saw no or minimal bubble formation.

If you add soap to water, however, you get lots of bubbles. This is because bubbles need just the right amount of surface tension to form. Plain water has too much, but soap decreases the surface tension of the water so that bubbles can form. If many bubbles are produced at the same time, they form a whole layer of bubbles on the water's surface that rises up inside the cup.

In contrast, when you blew through the straw into the cup with water and simethicone, even with detergent added, no significant bubble formation should have occurred. This is due to the fact that simethicone also decreases surface tension, but by a lot more than soap. If the surface tension of the gas bubbles gets too low, they will collapse which prevents bubble formation. The effect of simethicone increases the more you add to the water. With two simethicone tablets, you probably saw no bubbles form on the water surface at all whereas with only one-half of a tablet, bubbles might have been visible.

CLEANUP

Pour all the solutions down the sink and clean your work area with a wet paper towel. Wash your hands with water and soap.

Splitting Water

DID YOU KNOW THAT YOU CAN SPLIT WATER INTO ITS TWO COMPONENT ELEMENTS—WITH JUST A FEW HOUSEHOLD MATERIALS? SEE WHAT YOUR WATER IS REALLY MADE OF IN THIS MOLECULE-BUSTING ACTIVITY.

PROJECT TIME

60 Minutes

KEY CONCEPTS

Water
Chemistry
Electricity
Molecules

You use it every day and cannot survive without it—no it's not the internet, but water! It is one of the most essential compounds in the world and has become an increasingly important issue everywhere. You probably have heard that in many places droughts or water pollution limits clean drinking water supplies, and those supplies keep shrinking. When considering this, have you ever wondered why we do not just make our own water? What water is actually made of? In this science activity, you will use energy in the form of electricity to split water into its individual components. Gather all the materials and get ready to find out what they are—and separate them yourself!

BACKGROUND

To find out what water is made of, it helps to look at its chemical formula, which is H_2O. This basically tells us that the water molecule is composed of two elements: hydrogen and oxygen or, more precisely, two hydrogen atoms (H_2) and one oxygen atom (O). Hydrogen and oxygen are gases at room temperature. So, does this mean that we can just combine both gases and then we get water? It is not quite as simple as that. What is missing in the equation is the energy component of the chemical reaction. Making water from its elements produces a large amount of energy. Chemical reactions that produce energy are also called exothermic reactions. In the case of hydrogen and oxygen, the released energy is so high that it is almost impossible to control, and most of the time it leads to an explosion. Luckily, this reaction does not occur spontaneously and only happens when you ignite the gas mixture with a lighter.

If making water from its elements is so dangerous, what about the reverse reaction? Splitting water into its two components is much easier to do and is called water electrolysis. Making hydrogen or oxygen this way seems simple. But as you probably suspected, this reverse reaction needs an energy input, which is why it is also called an endothermic reaction. In water electrolysis, the source of energy used for the reaction to occur is electricity. An easy way to generate electricity is with a battery. Because pure water is not good at conducting electricity, however, electrolysis requires the addition of an electrolyte, such as salt or acid. The electrolyte dissolves in water and separates into ions (electrically charged particles) that move through the solutions and are able to conduct electricity this way. To add electricity to the solution, you also need two electrical conductors that make contact with the water. These are called electrodes and are mostly metals or other conductive materials. When electrical current is applied to the electrodes, the ions (electrically charged atoms) in the electrolyte—including the positively charged protons (H^+) and negatively charged hydroxyl ions (OH^-) that result from the self-ionization of water—start moving toward the electrode with the opposite charge where either hydrogen or oxygen is produced. You can see this for yourself in this activity and even capture both gases

MATERIALS

- Waterproof work area
- Adult helper
- Glue
- Mechanical pencil graphite (thick enough that it does not break)—or a wooden pencil and a knife with which to have your adult helper extract the graphite

- Play-Doh or other clay
- Two or three disposable plastic cups (Clear plastic 18-ounce cups work best.)
- Scissors
- Permanent marker
- Two metal thumbtacks
- Distilled water
- Baking soda
- 9-volt battery
- Two transparent straws (jumbo size)
- Teaspoon
- Medical dropper
- Timer
- Paper towels
- pH test strips (optional)
- Table salt (sodium chloride—NaCl) (optional)

PREPARATION

- Your work area should be waterproof; water spills might happen during the activity.
- Collect your sturdy mechanical pencil graphite. Make sure you have two pieces of about 1 inch (2.5 cm) in length. If you choose to use a wooden pencil, have an adult take the knife and extract the graphite lead from the pencil. (For this you should also get two pieces of about 1 inch [2.5 cm] in length). These will be your graphite electrodes.
- With the scissors, carefully cut (or have an adult cut) around the open end of one plastic cup so it is the height of the 9-volt battery. In the bottom of the cup, cut a hole so that the tip of the battery (with the two poles) can fit through.
- Take the other cup and hold the bottom of that cup on top of the 9-volt battery. With the permanent marker, on the inside of the cup draw two dots on the bottom where the cup touches the two poles of the battery.
- Take the cup away from the battery, and with a thumbtack, punch one hole through each of the points in the plastic cup bottom that you marked.

- Carefully insert the two graphite (pencil core) pins through the two holes, one each. You might need to make the holes a little bit bigger so they fit.

- Use the glue to make a seal around the graphite pins on the outside of the cup and let it dry. This should prevent water from leaking through. Make sure the graphite pins are not all covered in glue, otherwise they will not make contact with the battery anymore.

- Place the cut cup, open-side down, over the battery. The cup bottom and the battery should fit together, creating one level surface on which you can place the other cup.

- Put the cup with the graphite pins on the upside-down, cut cup, on top of the battery. It should sit firmly and each of the graphite pins should be in contact with one of the battery poles.

- Seal one end of both jumbo drinking straws with Play-Doh or clay.

PROCEDURE

- Take the cup with the graphite pins and pour approximately 1 1/4 cups (300 ml) of distilled water into the cup, away from the battery. Make sure that it is not leaking. If it is, you might need to add a bit more glue to make a tight seal. **Note:** You want to avoid touching the water or the electrodes once the cup is placed on top of the battery as you might feel the electricity tingling your fingers.

- As you did previously, place it on the upside-down, cut cup, on top of the battery so that each of the graphite pins makes contact with one of the battery poles. You might need to press it down a little to make a good connection. Observe the two graphite electrodes. *What do you see? Is anything happening at the electrodes?*

- Remove the cup filled with distilled water from the battery. Fill 1 teaspoon with baking soda and stir it into the distilled water until everything is dissolved. *What do you think the baking soda will change? What function does it have?*

- Now put the cup back on top of the battery and connect the graphite electrodes with the battery poles. *What do you observe now? Does anything happen at the graphite pins? What do you think the reaction's products are?* Compare the reactions that happen at each of the graphite electrodes. *Can you see a difference between both sides? Is there one graphite electrode at which the reaction is more pronounced? Which pole of the battery is this graphite pin connected to, the positive or negative?*

- Put your nose inside the cup and smell the reaction products. *Is there any smell? If so, how does it smell?*

- Remove the cup from the battery again. With the medicine dropper, fill both plugged-up jumbo straws with the baking soda solution from inside the cup that has the graphite pins. Once they are full, close each with one of your fingers and turn them upside down. Submerge them into the cup with the baking soda solution and carefully place them on top of the graphite pins (one straw on each) so that the straws stay completely filled with baking soda solution. If the straws do not stay upright, you can lean them against the side of the cup. *What do you think will happen with the straws?*

- Once the straws are placed on top of the graphite pins, put the cup back on top of the battery. Leave it there for 10 minutes and press the cup down a little to make sure that the electrodes stay connected and the electrode reactions are happening continuously throughout that time. Observe the jumbo straws that you put on top of the graphite pins. *What is happening to the water that you put in there? Do you notice a difference between the two water levels in both straws? Which one is higher, which is lower; to what battery poles are each of them connected?*

- After the 10 minutes are over, mark the water level in each of the straws with the permanent marker. *How much more water was displaced by the reaction products on the negative pole compared with the positive pole? Is it the same, double, or triple?*

EXTRA

If you have any pH strips that can measure the acidity or basicity of solutions, use them to measure the pH in each of the jumbo straws once the water level decreases by about 50 percent. Carefully remove the jumbo straws from the electrodes and immediately seal each one with a finger once you lift it off the electrodes. Making sure you do not lose the water that is inside, dip a pH test strip inside. *What color does the test strip show and what pH does this represent? Is there a difference between the solutions in the two straws? How do they differ, and why do you think this is the case?*

EXTRA

Repeat the experiment, but instead of adding baking soda to the distilled water, add a teaspoon of table salt (sodium chloride, or NaCl) and let the electrolysis run for five minutes. *Do the electrode reactions change? What about the smell of the reaction products; can you make out a certain smell this time? Why do you think this is?*

SCIENCE FAIR IDEA

Exchange the graphite electrodes with metal thumbtacks. You might need to use a fresh cup for this. Push the thumbtacks into the bottom of the cup so that they do not touch each other but so that each one touches one of the battery poles once you place the cup on top of the battery. With the thumbtacks no glue seal is needed. Repeat the original procedure but this time add 1 teaspoon of table salt to the distilled water. Observe the electrode reactions. *What happens this time?* Look closely at the pin that is connected to the positive battery pole. *Can you see other reaction products besides gas? What do you think happened? How do the metal thumbtacks look after you take them out again?*

OBSERVATIONS AND RESULTS

Were you able to split the water into hydrogen and oxygen gas? Did you see a lot of bubbles being produced at both of the graphite pins? Initially, when you put the cup with distilled water on top of the battery, you probably did not see much happening at the graphite electrodes. This is because distilled water does not conduct electricity very well. Therefore no or only minor electrode reactions are possible. If you add electrolytes such as baking soda, however, the added ions can conduct electricity and you should have seen gas bubbles emerging at both of the graphite pins. On one side, at the positive pole, oxygen develops whereas at the negative pole, hydrogen is produced.

When you looked closely, you might have observed that on the graphite electrode connected to the negative pole of the battery more gas was formed than on the other side. Collecting the two gases with the jumbo straws probably demonstrated this even better. After 10 minutes, the water level at the negative pole should have been about half that of the positive side, meaning that you collected about double the amount of hydrogen gas compared with oxygen. The difference is due to the fact that one water molecule has two hydrogen atoms to one oxygen atom, as explained above. This means it takes two water molecules ($2\ H_2O$) to make one oxygen molecule (O_2). At the same time, however, two molecules of water ($2\ H_2O$) can make two molecules of hydrogen ($2\ H_2$). Whereas hydrogen and oxygen are formed at the electrodes, the leftover reaction products from water are protons (H^+ on the side of oxygen) and hydroxyl ions (OH^- on the side of hydrogen). You can visualize this by putting a pH strip into the solutions in the jumbo straws above each electrode. The solution in the straw that was put on top of the negative battery pole electrode should show a basic pH (7 or greater) whereas the other one should be acidic (pH less than 7).

Other electrode reactions can happen, too, if there are ions in the solution that compete with the hydrogen or oxygen production. You might have observed once you added salt (sodium chloride) to your electrolyte, it started smelling like you were at a swimming pool. Instead of oxygen, chlorine is produced at the positive battery pole, which is also used for disinfection of swimming pool water. If you used metal thumbtacks as electrodes instead of graphite pins, the metal (usually steel or brass) will dissolve, or corrode, at the positive battery pole and instead of (or in addition to) gas production, you should see that the metal thumbtack turns reddish brown. This demonstrates that electrolysis is not only a technique for splitting water into its components but also can drive other reactions that otherwise would not occur spontaneously.

CLEANUP

Clean up any spills with a towel. Pour the water with the electrolyte (baking soda or salt) down the drain. Remove the electrodes (graphite pins or metal thumbtacks) from the cups and put them into the trash. Throw the plastic cups and the sealed jumbo straws into the trash. You can reuse the 9-volt battery.

THE SCIENTIFIC METHOD

The scientific method helps scientists—and students—gather facts to prove whether an idea is true. Using this method, scientists come up with ideas and then test those ideas by observing facts and drawing conclusions. You can use the scientific method to develop and test your own ideas.

Question: What do you want to learn? What problem needs to be solved? Be as specific as possible.
Research: Learn more about your topic and refine your question.
Hypothesis: Form an educated guess about what you think will answer your question. This allows you to make a prediction you can test.
Experiment: Create a test to learn if your hypothesis is correct. Limit the number of variables, or elements of the experiment that could change.
Analysis: Record your observations about the progress and results of your experiment. Then analyze your data to understand what it means.
Conclusion: Review all your data. Did the results of the experiment match the prediction? If so, your hypothesis was correct. If not, your hypothesis may need to be changed.

GLOSSARY

acidity: The level of acid in a substance.
compound: Composed of two or more separate elements.
concentration: The amount of a certain substance in a solution.
dissolve: To mix completely into a liquid.
decomposition: The state or process of decaying or breaking down.
dense: Packed very closely together.
hydrogen bond: A bond, or connection, between two hydrogen atoms caused by attraction between them.
ions: Any atom or group of atoms that has one or more positive or negative electrical charges.
mass: The amount of matter in an object.
molecule: A group of atoms bonded together.
organ: A part of the body that performs a specific function.
oxidation: When oxygen combines with an element or compound.
polar: A structure having an uneven distribution of electrons.
reaction: Resistance exerted in opposition to the impact or pressure of something.
solution: A liquid mixture in which the minor component (the solute) is equally distributed within the major component (the solvent).
substance: A particular kind of matter with the same properties.
transparent: Allowing light to pass through so that objects behind can be distinctly seen.

ADDITIONAL RESOURCES

Books

Dickinson, Michelle. *Kitchen Cabinet Science Projects: Fifty Amazing Science Experiments to Make with Everyday Ingredients*. New York, NY: Penguin Workshop, an imprint of Penguin Random House LLC, 2021.

Heinecke, Liz Lee. *Physics for Kids: Science Experiments and Activities Inspired By Awesome Physicists, Past and Present*. Beverly, MA: Quarto Publishing Group USA Inc., 2022.

Heinecke, Liz Lee. *Sheet Pan Science: 25 Fun, Simple Science Experiments for the Kitchen Table*. Beverly, MA: Quarry Books, 2022.

Websites

Discovery Education
sciencefaircentral.com

Exploratorium
www.exploratorium.edu/search/science%20fair%20projects

Science Buddies
www.sciencebuddies.org/science-fair-projects/project-ideas/list

Science Fun for Everyone!
www.sciencefun.org/?s=science+fair

Videos

Bubble Wand Experiment and Science Journal | Let's Learn
https://ny.pbslearningmedia.org/resource/ll33-bubble-wand-video/lets-learn/

Water Treatment Plant
https://ny.pbslearningmedia.org/resource/ess05.sci.ess.watcyc.h2otreatment/water-treatment-plant/

Publisher's note to educators and parents: Our editors have carefully reviewed these websites to ensure that they are suitable for students. Many websites change frequently, however, and we cannot guarantee that a site's future contents will continue to meet our high standards of quality and educational value. Be advised that students should be closely supervised whenever they access the internet.

INDEX